SPACE STATION

NECIA H. APFEL

SPACE STATION

FRANKLIN WATTS I A FIRST BOOK
NEW YORK I LONDON I TORONTO I SYDNEY I 1987

Diagrams by Vantage Art

Cover photograph courtesy of NASA

All photographs courtesy of NASA
except Smithsonian Institution,
National Air & Space Museum: p. 12;
Hughes Aircraft Company: p.22

Library of Congress Cataloging-in-Publication Data

Apfel, Necia H.
Space station.

(A First book)
Bibliography: p.
Includes index.
Summary: Discusses the achievements of outer
space exploration in this century and examines
future possibilities such as a permanent
space station, colonies in space, and journeys
outside the solar system.
1. Outer space—Exploration—Juvenile literature.
2. Space colonies—Juvenile literature. [1. Outer
space—Exploration. 2. Space stations] I. Title
TL793.A63 1987   910'.0919   87-2217
ISBN 0-531-10394-3

CONTENTS

CHAPTER ONE
BLAST OFF
INTO SPACE

"10 . . . 9 . . . 8 . . . 7 . . ."

You are lying on your back, securely strapped into your seat, about 80 feet (24 m) above the ground, awaiting lift-off. You hear an enormous growl from far below as the main engines ignite. The spacecraft begins to vibrate as the rocket boosters build up more and more thrust.

"6 . . . 5 . . . 4 . . ."

The noise and vibration grow louder. Even with the protective helmet you are wearing, the sound is very great.

"3 . . . 2 . . . 1 . . ."

"We have ignition . . ."

"We have lift-off . . ."

"All systems are go . . ."

You may not hear these words from your position within the spacecraft, but you will feel the surge of motion. The service tower that stood beside the craft quickly drops from sight, and within seconds you are aware of the accelerating force of 3gs.* You are rapidly headed for space.

---

*"3gs" is three times the force of the earth's gravity. The faster you accelerate, the greater you feel this force. If you could weigh yourself while accelerating at 3gs, you would find that the scale registers your weight as three times more than what you normally weigh. At 3gs your body feels very heavy, and any motion on your part under such conditions is very difficult.

Thirty seconds into the flight, you notice that the blue sky has become jet black, much blacker than the nighttime sky back on earth. You are above the atmosphere, 28 miles (45 km) up, hurtling through space at 3,000 miles (4,830 km) per hour. But the acceleration and the ascent continue.

Then, after you have been traveling for about eight minutes, your ride becomes much quieter and smoother. Your spacecraft is now moving at over 17,000 miles (27,370 km) per hour, some 200 miles (322 km) above the earth's surface.

You notice that the feeling of heaviness has left, and in its place is an unusual sensation. You cautiously take a pen from your shirt pocket and release it in midair. It does not fall, but instead hovers right in front of you. Now you understand the new feeling—you and the pen are weightless. You are now in zero g.

When finally you can unstrap yourself from your seat, you float out into the cabin. If you relax, your arms will float out from your body without any effort on your part. A slight push off one wall will send you sailing across the cabin. What fun!

Very few people have experienced weightless space flight, but within the next fifty years, it may become as common an activity as airplane trips are today. If this is an adventure you look forward to, you are not alone. The desire to go out into space is a dream that has existed in some form for as long as humans have recorded their ideas and maybe even longer. The exploration of space can be compared to the opening of the American West about two hundred years ago or the exploration of Antarctica at the beginning of this century.

What was probably the first fantasy space trip was described by the Greek writer, Lucian of Samosata, in the third century A.D. In his *Alethes Historia* (True History) he tells the story of a sailing ship that gets caught in a great wind and is blown into the sky. The boat travels through space, lands on another world which Lucian describes, and then it returns to earth. Although Lucian warns his readers that the story is not true, his tale certainly must have had great influence on science-fiction writers living centuries later.

In the seventeenth century, the great German scientist Johannes Kepler wrote *Somnium* (Dream), a most unusual science-fiction story telling of a trip to the moon. Using the best astronomical knowledge of his day plus an obviously vivid imagination, Kepler described the moon. Since Kepler was primarily a scientist, it has been suggested that he may have used the science-fiction format to mask controversial ideas about the moon and outer space that he and other scientists had at that time.

Another fascinating story of space travel, this one from the eighteenth century, is *Micromegas*, written by one of the greatest French authors, Voltaire. His hero, Micromegas, comes from the large, bright star Sirius where everything is gigantic compared to things on earth. Micromegas sets out to explore the universe and eventually finds our solar system. He is not impressed although he recognizes that intelligent life does exist on earth. Voltaire, in this way, was trying to point out the insignificance of man and his planet on the cosmic scale.

Perhaps the most famous fantasy space trip was the one described by Jules Verne in 1865. *From the Earth to the Moon* is remarkable because, although by our scientific standards it has many faults, it anticipated much of today's space exploration. Scientists and science-fiction readers still read, admire, and enjoy the writings of Jules Verne.

A few years after the publication of Verne's moon fantasy, a Boston clergyman, the Rev. Edward Everett Hale, wrote a fictitious series of articles about a huge spherical space station built of bricks and whitewashed on its outside so it could be seen more easily in the sky. It was to be used as an aid to navigation. Published in the *Atlantic Monthly* magazine in 1869, the story was appropriately called "The Brick Moon."

One problem that troubled all of these writers of fantasy was devising a suitable method to propel their spaceships into space. As we saw, Lucian of Samosata used a tremendous wind. Kepler dreamed up some demons who could only fly through space during the night. Voltaire's hero, Micromegas, used sunbeams, comets, and gravitation. Verne imagined an enormous 900-foot (274–m) cannon, and Hale used a gigantic flywheel driven by a waterfall to get his brick "moon" into space. Even as recently as the beginning of the twentieth century, science-fiction writer H. G. Wells invented a special antigravity material called Cavorite to propel his spaceship in *The First Men in the Moon*. But by then the early technology of rocketry was beginning and what had been only a dream was soon to become reality.

CHAPTER TWO

THE DAWN OF THE SPACE AGE

When you throw a ball high into the air, you know that it will stop traveling upward very quickly and start a downward path instead. The harder you throw the ball, the faster and higher the ball goes and the longer it takes for it to return to earth. Wouldn't it be exciting if you could toss a ball so high it would not return but instead would go out into space? Even dreamers of space travel knew that the strongest person in the world could never throw an object with enough force to achieve that goal, but certainly sunbeams, demons, or flywheels wouldn't be much better.

But these early science fiction writers knew that to get into space, a means of propulsion strong enough to counteract earth's gravity was required. To achieve this propulsion strength, space enthusiasts turned to the field of rocketry. Long ago they had discovered that pointed objects, rather than spherical or flat ones, encountered much less air resistance as they were propelled upward.

The early rockets at first received their propulsion power from gunpowder. Later, however, less dangerous but more powerful propellants were used. Also, the first rockets were not designed to lift an object into outer space but rather were used as weapons during warfare or as firecrackers to celebrate some festivity.

DEADLY FIRECRACKERS

The Chinese are usually given credit for inventing the rocket as well as discovering black powder, which, when guns were invented,

Early Chinese fire arrow rockets

became known as gunpowder. Black powder dates back two thousand years, but no one is sure when the first rocket was built. However, by the thirteenth century, powder-propelled fire arrows were used in China with devastating effects on the enemy. The military use of these and other rocket-type devices spread quickly to the Arabic and European nations as military men saw their obvious advantages in battle.

Although some other uses for rockets were also found in the centuries that followed, the interest in them generally rose during wartime. This was one of the reasons that research in rocketry declined in the latter part of the nineteenth century as the world found itself in an unaccustomed period of peace. Even during World War I, however, the rocket lost its importance because it could not compete with the much greater accuracy of the new artillery weapons.

In the late 1890s, Pedro A. Paulet, a Peruvian chemical engineer who was one of the few rocket researchers of that period, discovered a liquid propellant that could replace the highly explosive and relatively uncontrollable black powder. And at the same time Konstantin Tsiolkovsky, a Russian school teacher and scientist, began to work out the basic principles of space flight using rockets. Later, in 1911, Tsiolkovsky wrote a science fiction article for a Russian magazine describing the observation of the earth from a rocketship whose orbital period around the Earth was two hours. But he knew, of course, that before such a spacecraft could become a reality, very powerful rockets were needed.

NEWTON'S THIRD LAW

How does a rocket work? To understand the underlying concept of rockets, we must look first at the three fundamental laws of motion propounded by the famous English scientist Sir Isaac Newton in the seventeenth century:

1. *A body remains at rest or in uniform motion in a straight line unless acted upon by a force.*
2. *Force equals mass multiplied by acceleration (F = ma).*
3. *To every action there is an equal and opposite reaction.*

Newton's third law best helps to explain rocketry, although the law can be illustrated in other ways. For example, astronauts moving about in their weightless, orbiting spaceships demonstrate the action-reaction principle beautifully. The slightest push they give in one direction will send them flying off in the opposite direction. Because no other forces are pulling on them, nothing can stop their flight except the cabin wall on the opposite side. Here, the earth's gravity pulls us downward so that light pushes against a wall will do no more than possibly upset our balance. However, if you are in a rowboat and attempt to step out of it to get up onto the dock, your push on the boat as you step upward may very well send the boat floating back out into the water again.

It is easy to demonstrate the principle of action-reaction right now if you have a toy balloon. Blow up the balloon but don't tie off the opening. Now release the balloon. What happens? The pressure of the air in the balloon rushing out of the hole pushes the balloon in the opposite direction. Although rockets and jet airplanes are much better controlled than the wildly careening balloon, they both operate on the same principle.

In a rocket, the propellant is made to produce a hot gas which rushes out through the exhaust, creating enough thrust to move the rocket. In a jet airplane the combustion of its fuel with the help of the oxygen in the air creates the gas pressure. In other words, an airplane needs the air around it to burn its fuel and thus create thrust. A rocket, on the other hand, creates its own hot gas from the interaction of the chemicals used.

Although the Russian scientist Tsiolkovsky never constructed a rocket himself, he was able to calculate the power such a device would have and the power that was needed to enable it to go off into

Newton's third law, "to every action there is an equal and opposite reaction," can be illustrated in the following way: As you step out of a rowboat and onto the dock, your push on the boat as you step upward sends the boat floating back out into the water again.

space. He knew that although the liquid propellants discovered by Paulet could send a rocket a great distance, they were not powerful enough to enable a rocket to be shot into space. Tsiolkovsky had calculated that the strength of the earth's gravitational attraction was enough to pull such rockets back to the ground just as the same gravitational attraction pulls any ball down that you throw up into the sky.

Tsiolkovsky proposed instead the building of a series of rockets mounted one on top of the other. As each rocket stage used up its propellant, it would separate from the main body and fall back to the ground. Thus, the later stages would be given a running start, enabling them to climb much higher than the first stage. As you watch the launching of a spacecraft, you will see how Tsiolkovsky's ideas have been put to work.

THE FATHER OF
MODERN ROCKETRY

The Russians consider Tsiolkovsky "the father of astronautics." But since he never worked on any rockets, the title of "father of modern rocketry" must go the American scientist and inventor Robert Hutchings Goddard. Goddard and Tsiolkovsky never knew of each other's work but independently came to similar conclusions about rockets and rocket propulsion. Goddard also developed the concept of a multistage rocket using a liquid propellant, but he experimented and was granted patents for his rocket inventions.

In 1926, Goddard fired his first liquid-fuel rocket, which reached a speed of 60 miles (97 km) per hour and an altitude of 41 feet (12 m). That flight lasted only 2.5 seconds. By 1937, Goddard's rocket design had been streamlined and a test model was able to reach 3,250 feet (991 m) in a flight that lasted 29.5 seconds. That

Robert Hutchings Goddard launched the first liquid-propellant rocket in 1926. When the United States began to prepare for the conquest of space in the 1950s, American rocket scientists began to recognize the debt owed to Goddard. They discovered that it was virtually impossible to construct a rocket or launch a satellite without acknowledging his pioneer work.

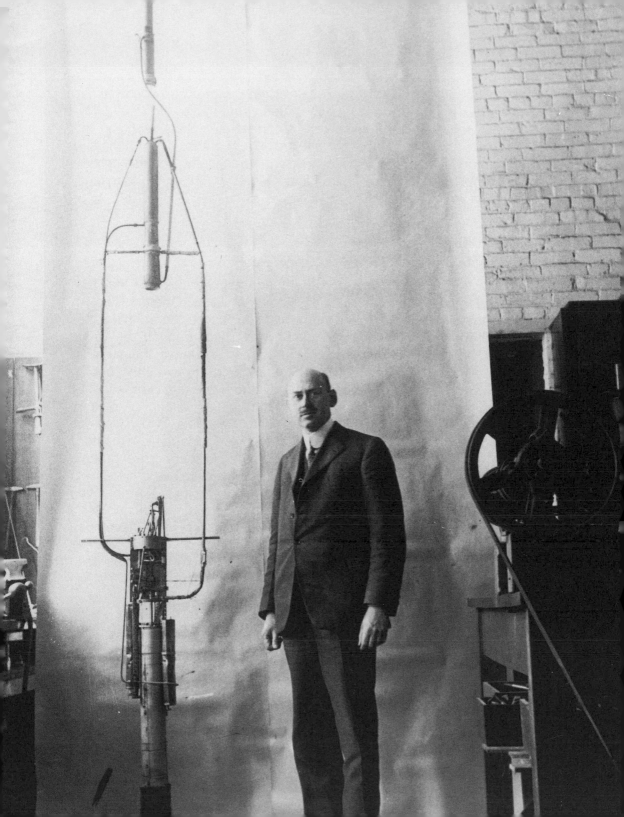

rocket was almost 18 feet (5.5 m) long, more than twenty times longer than the one launched eleven years earlier.

Goddard, like Tsiolkovsky, dreamed of rockets reaching the moon and outer space, but with the advent of World War II, he had to turn his attention to the war effort. In recognition of his achievements, the Goddard Space Flight Center at Greenbelt, Maryland, was named after him.

THE GERMAN
ROCKET ENGINEERS

The third great pioneer of rocketry and astronautics was the Hungarian-born German scientist Hermann Julius Oberth. He also wanted to design rockets that would soar above the earth's atmosphere and reach the moon and beyond. But throughout the 1920s he received little encouragement from any academic, industrial, or governmental source for his rocket research. He wrote two books, however, that received wide publicity and made him internationally famous. In them he discussed just about every aspect of rocketry and space flight. These inspired other scientists to continue his work and ultimately led to many of today's space achievements.

Like Goddard, Oberth had to put aside his space-travel dreams and use his talents for his government. In his case this was the Nazi regime. He worked with the German rocket expert Wernher von Braun at Peenemunde, the German rocket development center on the Baltic coast where the infamous V-2 rockets were developed and constructed. The V-2, a liquid-propellant rocket, was designed to bomb Britain, not to reach outer space, but its design ushered in a new phase in rocketry. It has been called the true ancestor of the modern spaceship.

After World War II, the German scientists who worked on the V-2 and other rocket designs found themselves in great demand by both the United States and the Soviet Union. Most of them were able to come to the United States and, with the help of American scientists, soon established a proving ground at White Sands in

New Mexico. There they first tested undamaged, captured V-2 rockets and then proceeded to forge ahead in the development of bigger and more powerful rockets. Then a rocket site was set up at Cape Canaveral, on the eastern coast of Florida. Practically all the major space shots were launched from there.

Although the growth of rocketry in the Soviet Union was kept secret by that country, it is assumed that their program was similar to that of the United States. Using the technological knowledge and experience of the German rocket engineers, as well as the research of their own scientists in the postwar years, the United States and the Soviet Union found themselves constantly competing with each other for rocketry superiority. And as the power and reliability of the rockets improved, the rocket enthusiasts in both countries turned their attention back to their original objective—space. But for that they would need more than rockets. They would need spacecraft.

CHAPTER THREE

THE FIRST SPACE PROBES

The story of flight, whether in the earth's atmosphere or above it, would not be complete without mentioning the historic event that occurred on December 17, 1903. On that day Orville Wright and his brother, Wilbur, two American inventors and aviation pioneers, were the first to successfully fly a power-driven aircraft. Their first flight was no longer than the length of a large airplane today, but by the fourth and final flight of the day they had managed to fly 852 feet (260 m) in fifty-nine seconds. Little did they imagine that only sixty-six years later the first human would walk on the surface of the moon. That development and what lies beyond it is our subject matter now.

THE RACE TO SPACE

In 1955, President Dwight D. Eisenhower announced plans to launch a small unmanned artificial earth satellite sometime in late 1957 or 1958. At the same time and with great secrecy, the Russians were already designing and building their first satellite. They took the world by surprise when *Sputnik I* was launched on October 4, 1957. And before the United States could launch its own satellite, *Explorer I*, the Russians sent up *Sputnik II* with a dog named Laika aboard as a passenger. Weighing 1,120 pounds (508 kg), *Sputnik II* also carried many instruments to gather information about the upper atmosphere.

Explorer I went into orbit on January 31, 1958. It was the beginning of a long series of successful Explorer and later Vanguard satellite launches by the United States. Although the Soviet programs were more spectacular at the beginning, the United States gathered more knowledge about the earth and its immediate surroundings. Unmanned satellites that aided in navigation, weather prediction, and communications were employed in this new method of exploration.

Once the technology of earth-orbiting satellites had been achieved, scientists and engineers turned some of their attention to the regions beyond the earth. The moon, because it is the closest body to the earth, became their first target. Once again, the Soviets were first with their *Luna I* spacecraft that was launched January 2, 1959. It flew within 5,000 miles (8,050 km) of the lunar surface before continuing out into space. Then, in September 1959, the Russian *Luna 2* spacecraft reached the moon. It crash-landed on the lunar surface but not before sending back information and photographs of the moon. It was the first artificial object to land on another celestial body. Less than a month later another Russian probe, *Luna 3*, circled the moon and took the first pictures of its far side, the side that we never see here on earth.

The American lunar exploration program got off to a slower start, but by 1965 several of the Ranger spacecraft had sent back to earth thousands of pictures of the lunar surface. Then, on May 30, 1966, the American Surveyor spacecraft made a soft landing on the moon. The Russians had by that time also achieved a lunar soft landing, but the pictures and other data collected by the Ranger and Surveyor series of spacecraft later proved vital to future lunar probes. Between 1969 and 1972, the United States successfully launched six manned Apollo misions to the moon. Each three-man crew landed on the moon, explored it, and then returned to earth.

During the 1960s and 1970s, American and Russian unmanned spacecraft also reached the planets, photographing all of them as far away as Jupiter. Saturn and Uranus were reached in the 1980s, and Neptune will be visited next, in 1989. Russian spacecraft land-

ed capsules on Venus and Mars while the American *Viking I* and *Viking II* spacecraft landed and photographed the Martian landscape.

All of these launches relied upon multistage rockets, which were much more powerful than any designed by Tsiolkovsky, Goddard, or Oberth. But the original concept that these three pioneers had developed was still being used successfully.

BUT HOW DOES A ROCKET WORK?

Before we go further in our story of spacecraft, it is important to understand some of the theories involved in rocket launchings, orbiting, and travels through space. How much power is necessary to launch a spacecraft? Once up there, what keeps it from falling back to the ground? Why does it stay in orbit? How can it be sent into deeper space, to the moon or planets?

In the seventeenth century, the famous Italian scientist Galileo dropped two objects of different weights from the top of the Leaning Tower of Pisa. They both landed at the same time, proving that all bodies here on earth, regardless of their masses, fall at the same rate. From many experiments since then we know that here on earth all bodies in a vacuum fall approximately 16 feet (5 m) in the first second of their descent.

The reason we measure the rate of descent in a vacuum is to remove any air resistance from our calculations. You know that a leaf, for example, does not fall as fast as an acorn but that is because the leaf floats gradually down to the ground, pushed around by air currents, whereas the acorn, because of its shape,

In this historic photo, an Apollo 12 *astronaut approaches a Surveyor spacecraft on the moon in November of 1969. The Surveyor spacecrafts had paved the way for the Apollo landings.*

falls straight down and is affected much less by the air currents. In a vacuum, both the leaf and acorn would fall at the same rate.

The earth, as you know, is round, and measurements show that it curves away from a straight line about 16 feet (5 m) every 5 miles (8 km). Now imagine an earth with a surface that is precisely circular and completely smooth, with no mountains or valleys or atmosphere to contend with. Let us build a huge tower on this mythical earth and fire a rocket from the top of it so that it travels on a path that is parallel to the ground. We give the rocket just enough power to enable it to move at a velocity of 5 miles per second.

In the first second, the rocket zooms rapidly over 5 miles of land. But in that same first second, just like any other falling body on earth, the rocket falls 16 feet. However, the curvature of the earth, which is 16 feet for every 5 miles, causes the earth's surface to curve away from the rocket by the same amount as its fall, and so the rocket never reaches the ground or comes any closer to it. This same thing happens during the next second and the one after that, and so forth.

As you can see in the diagram, the rocket will never land but will remain in orbit just above the earth's surface. Its rate of fall will equal the amount of the earth's curvature, so that, in a sense, the surface will continually "move" away from the rocket. In the same way, the moon in its orbit around the earth is continually "falling" toward the earth.

If we decrease our imaginary rocket's velocity, it will eventually fall to the ground. That is because it will still be falling at 16 feet per second but not moving fast enough across the land for the earth to curve away from it. However, until the rocket hits the surface, it is in orbit. This is also true for a thrown baseball or for a bullet ejected from a gun. Neither of these travel at 5 miles per second, but if in some way the earth's mass was instantaneously compressed into a tiny ball just after the baseball was thrown so that the ball didn't hit the ground, it would continue to orbit the tiny earth.

What happens if we increase our rocket's velocity to more than 5 miles per second? Then its orbit becomes elongated or elliptical

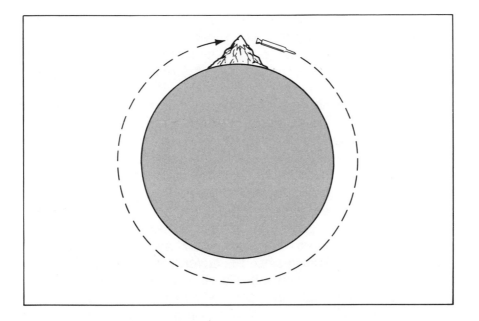

Imagine firing a rocket from the top of a very high mountain. The rocket will be subjected to the pull of the earth and will follow a curved path. Conceivably, it could be fired with enough initial speed so that it would circle the earth like a moon.

rather than circular. It still returns to its starting point on top of the tower, but, depending upon how much above the 5-miles-per-second velocity it receives, its orbit on the other side of the earth extends farther and farther away. Note, however, that it continues to go around the earth.

Thus, with each increase in velocity, the rocket's orbit becomes more and more elongated until it extends out beyond the moon itself. Finally, if the velocity is increased to 7 miles (11 km) per second, the rocket's orbit changes from elliptical to parabolic. A parabolic orbit does not go around the earth. It is open-ended and never returns to its starting point. The rocket thus escapes the earth's gravitational pull and goes out into space. Seven miles (11.2 km) per second is called the "escape velocity" of the earth. It is equal to 25,000 miles (40,250 km) per hour.

The required orbital or escape velocity for a spacecraft can therefore be calculated with great accuracy, depending upon its desired destination. Of course, heavier spacecraft need more powerful rockets to reach such velocities, which is why the multistage rocket boosters have been so successful. But once the craft reaches its orbiting velocity around the earth, it and everything in it are weightless. Then, extremely small booster rockets can be used to gently nudge the spacecraft into proper position.

The theory of how to send rockets and spacecraft into orbit and beyond has been known since the seventeenth century. It was Sir Isaac Newton who, along with his three laws of motion, also set forth a law of universal gravitation and spoke of the possibility of an earth-orbiting satellite. From Newton's laws of motion and gravitation came the equations that enable scientists even today to compute velocities, orbit sizes, and other necessary information. Although Newton could compute all of this, the technology of rocketry did not come until the twentieth century. Until then, all dreams of exploring outer space had to remain solely on paper.

CHAPTER FOUR
MANNED SPACE FLIGHT

An enormous amount of courage is needed to go where no human has ever ventured before. The first astronauts and cosmonauts demonstrated this courage for, although many test animals had already been sent aloft with no ill effects, no one knew what the effects of space travel might be on the human body.

Major Yuri A. Gagarin, a Russian cosmonaut, was the first to attempt this feat. His one-orbit flight around the earth in a small spacecraft called *Vostok 1* took place on April 12, 1961. His trip lasted all of one hour and forty-eight minutes. Later that year, Major Gherman S. Titov, another Russian cosmonaut, stayed in orbit for seventeen orbits in *Vostok 2.* His journey took a little over twenty-five hours.

The first American to go into orbit was Marine Colonel John H. Glenn, Jr. His flight took place in a Mercury spacecraft called *Friendship 7* on February 20, 1962. Glenn's craft stayed aloft for almost five hours, making three orbits of the earth in that time. In the previous year, two other Americans, Commander Alan B. Shepard and Major Virgil I. Grissom, had both made suborbital flights to test the Mercury spacecraft. These and other one-person flights in the early 1960s conclusively proved that humans could go into space.

The endurance record in space quickly grew from hours to days to weeks and even to months. At the same time, the size of the craft was enlarged to permit two-person and three-person crews aboard. Tests to demonstrate that these spacecraft could rendezvous and even dock with each other in space were conducted by both the United States and the Soviet Union. There was even an international docking of an American and Russian craft. It was called the Apollo-Soyuz mission. The astronauts and cosmonauts visited each other in space before separating their ships and finishing their flights.

"Spacewalks" outside their spacecraft were taken by several astronauts and cosmonauts, proving the ability of humans to leave their spacecraft in properly designed spacesuits. Although tethered to the craft in the early spacewalks, the spacewalker had complete freedom of movement and was able to maneuver himself with the aid of a hand-held gas gun. These exercises showed that it was possible to perform tasks on the spacecraft's exterior, including repairs and replacements of film or batteries.

In the very earliest flights crew members remained strapped into their seats during the entire time they were aloft. Once the space flights became more than a few hours long, however, it was necessary to increase the size of the craft to allow the crew to move around. Nevertheless, even the Apollo spacecraft with its three-person crew was very crowded and little space was available. Although the command module in which the three men spent most of their time was nearly 11 feet (3 m) high and 13 feet (4 m) in diameter, the crew compartment within the command module had only 210 cubic feet (6 cu m) of living space. Because the trips to the moon took several days each way, one important criterion in choosing the Apollo crews was their ability to live with each other in such cramped quarters.

THE FIRST SPACE STATION

Five months after the end of the Apollo program in December 1972, the United States launched its first space station, Skylab. The Soviet Union also had its space station, Salyut, in orbit at about the

*Astronaut Owen Garriott, one of a crew of three
who lived on board Skylab for fifty-nine days in 1973.
Garriott is shown here at the Apollo Telescope Mount
console; from this console he made solar observations.*

same time. We know very little about the Russian craft, but the American one was quite different from any previous spacecraft. For one thing, Skylab was first launched and put into orbit without a crew. Then three-person crews were sent up successively to work in this orbiting space laboratory. Each of the crews successfuly docked its spacecraft to Skylab, entered it, and were able to conduct many experiments before returning to earth, leaving Skylab for the next visitors. The crew lived in this primitive space station for periods ranging from one month to almost three months.

The record for the longest stay in space, however, goes to the Russians who had a Salyut crew stay in their space station for more than six months. As the scientists from both the United States and the Soviet Union discovered, these long periods in space produced definite physical changes in the human body. These changes do not appear to be permanent ones, but doctors of space medicine are still unsure whether longer space trips will permanently endanger a healthy body.

LIVING IN ZERO GRAVITY

Once you are out in space, the first change in your body is quite noticeable and occurs rather quickly. Your body fluids redistribute themselves because they are free from the pull of earth's gravity. Instead of being forced downward toward your feet, they tend to flow toward your head. This causes your face to puff out a little and the proportions of the rest of your body to change. An astronaut can lose as much as four inches (10 cm) around his or her waist, but at the same time gain an inch or two in height. This increase in length happens more slowly than the puffy face change but is also due to the removal of gravity that compresses the disks in all of our spines here on earth. Of course, the astronauts return to their former dimensions soon after returning to earth.

More serious effects of prolonged weightlessness are the loss of calcium in the bones and the loss of some muscle capacity. Some astronauts have difficulty at first readjusting to gravity once they are back on earth. Exercise machines, including treadmills and stationary bicycles, were part of the equipment on Skylab and on other later spacecraft to enable the astronauts to stay in good physical condition. These helped to maintain body tone, but the effects of being in space for more than a week or so were noticeable.

Another physical problem that has plagued at least fifty percent of all astronauts is motion sickness. The sensitive organs of balance, located within our inner ears, are the first to be affected by zero gravity. They tell us when we are moving faster and in what direction we are going. They keep us oriented here on earth so that we know whether we are standing up, lying down, or possibly upside down. But in space these directions are meaningless and it is easy to become disoriented.

Not everyone suffers from motion sickness because of this disorientation, and doctors who specialize in space medicine have not yet been able to predict who will be sick. None of the tests given here on earth seems able to detect those who will suffer in space. Luckily, there are medications to relieve the symptoms, and the condition itself disappears fairly quickly and does not return.

Because of the return of the first astronauts in relatively good health and because of the many tasks they proved capable of performing while in orbit, it was felt that the value of humans in space had been demonstrated. Many scientists and engineers now felt safe in proposing a long-term space station in which many more people could live for longer periods of time. They wanted a space station that would be permanent even if the people who were sent to it did not remain there more than a few months. Because it would be staffed with far more than just three astronauts, a continuous flow of personnel back and forth could be achieved.

The cost of all of these space probes, however, was very expensive. The spacecrafts and the rockets that send them into space usually could only be used once, and the part that finally is retrieved is extremely small compared to the original equipment launched. For example, the vehicle that stood on the launching pad to send the Apollo astronauts to the moon was over 360 feet (110

m) tall. But only the 22-foot (7-m) cylinder that brought the astronauts back home survived. Just think of how expensive it would be if you had to buy a new bicycle every time you went to school.

THE SPACE SHUTTLE

What was needed first was a spacecraft that could be used repeatedly. When this was realized, the concept of the space shuttle began to evolve. It was already being designed by 1976. Although it is not a space station, it is the vehicle that will bring the equipment and construction material and crews into space to build the station. And it will eventually ferry the many scientists, engineers, and other personnel to and from the finished station.

Designed to take off like a rocket, stay in orbit like a spacecraft, and land like an airplane, it saves enormous sums of money by being reusable, and it has enabled people who are not trained astronauts to enter space. Besides the many experiments conducted by those aboard the shuttle, it has also launched many different satellites from its large cargo bay. These include communication, weather, and military satellites that would have had to be launched separately from the ground if it were not for the shuttle.

The shuttle also has put into orbit small scientific spacelabs, and many more are planned for future flights. These include the 94-inch (2.4-m) space telescope, the Galileo space probe to Jupiter, and the Ulysses craft that will fly over the sun's polar regions. All three are ready for launching. Other space probes, such as the Magellan craft that will be sent to Venus, are currently being built or designed.

A FLOATING REPAIR SHOP

The value of the shuttle and the advantages of having people working in space were shown when astronauts aboard the shuttle demonstrated their ability to retrieve and repair satellites that had malfunctioned. Retrieval of these satellites was accomplished with

*This cutaway view of the space shuttle shows
the crew at their duty stations on the flight deck.
The shuttle's payload bay doors are open.*

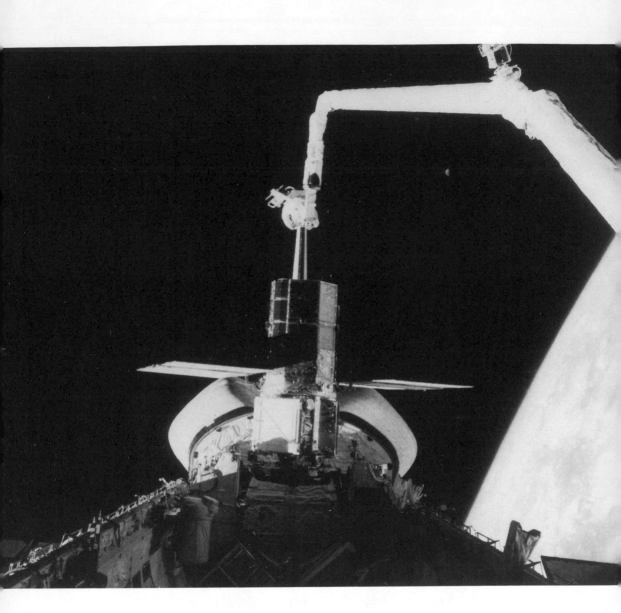

*The Canadian ''remote manipulator system''
being used to repair an artificial satellite*

the use of a 50-foot (15-m) manipulator arm that was made in Canada. It is attached to the cargo bay of the shuttle and is operated by remote control from within the shuttle cabin. Although it is made of metal, its design is patterned after the human arm with its shoulder, elbow, and wrist joints.

This long metal arm, which is called the "remote manipulator system," can be raised out of the cargo bay once the shuttle is in orbit and the bay doors are opened. After the arm has retrieved the malfunctioning satellite, and if it cannot be fixed while on board the shuttle, the satellite is placed in the cargo bay and returned to earth for repairs. It can then be relaunched by later space shuttle missions.

LIFE IN A SHUTTLE

Although the space shuttle is not a space station, it is one of the steps toward such a goal. In fact, many of the projects that take place aboard the shuttle are similar to those that may eventually be worked on aboard a space station. But the shuttle was not designed for long periods of habitation; most flights have only been for a week or less. Therefore, the living quarters for the crew are relatively small, several times bigger than that of the Apollo spacecraft but much smaller than that of Skylab. And whereas Skylab had four rooms composing its crew's living space, the shuttle's less elaborate construction has only one. If more than the minimum crew of two or three are aboard, as they have been in all but the very early flights, sleeping, eating, and working must be done in shifts to avoid getting in each other's way.

The cockpit of the shuttle resembles that of a conventional airliner, although it does have a spacecraft's guidance indicator that gives the craft's orientation and motion in space. Also, its computer system, into which all the other instruments are tied, is much more complicated than any airliner's system. There are four computers aboard the shuttle and all are constantly checking on the readings of the other three to lessen the possibility of any malfunction. The

35—

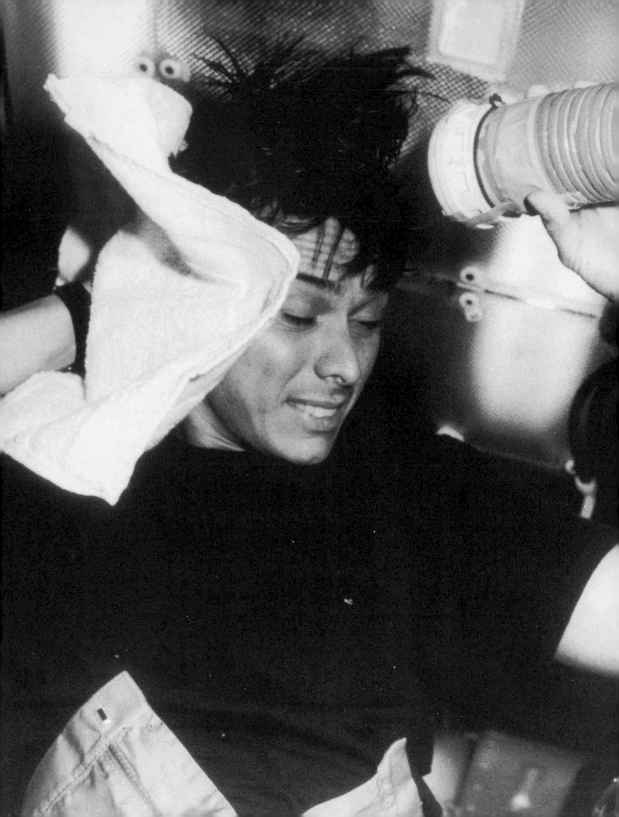

shuttle is also much more automated than any past spacecraft or aircraft, and the computer system is capable of flying it from space to the ground without human assistance.

Once all four shuttles had made their maiden voyages, the criteria for picking crew members were greatly relaxed. Although there were still two or three trained astronauts aboard every flight, the rest of the people have been scientists and engineers in various fields, military personnel, several U.S. senators, a high school teacher, and a few dignitaries from foreign countries. The foreign dignitaries usually were on board because their countries were launching a communication satellite on that mission.

A sad postscript to the history of the shuttle must be added here. In spite of all the careful precautions taken to ensure the safety of the shuttle, on January 28, 1986, the shuttle *Challenger* exploded two minutes after its launch, killing all seven people aboard. Besides the tragic loss of life and the destruction of one of only four existing shuttles, the mishap set the entire American space program back at least a year and possibly more. Nevertheless, with time, the entry of humans into space will continue once again.

The shuttle has already given more than a hint of what could be accomplished with a real space station. In his State of the Union message to Congress on January 25, 1984, President Reagan told of his decision to direct the National Aeronautics and Space Administration (NASA) to develop a permanently manned space station within a decade. It now may take longer than the ten years allowed, but the dreams of centuries have definitely taken another major step toward reality.

Astronaut Frank R. Chang-Diaz, mission specialist on board the Columbia, *uses some air from the space shuttle's airlock to dry his hair following a shampoo.*

CHAPTER FIVE

THE 1990S SPACE STATION

How do we build a permanent space station? To be worthwhile it should be much larger and far more elaborate than either Skylab or Salyut spacecraft, neither of which was meant to last for any length of time or to house a crew for more than a few months at a time. But those first primitive stations were completely built here on earth and then sent into orbit. The kind of space station that is now being proposed is too large and massive to be launched in one piece. It will have to be assembled in space.

Many of the parts or "modules" of the future station, however, will be built on earth and then carried into space by the space shuttle. Because the cargo bay of the shuttle is only 60 feet (19 m) long, none of these modules can be larger than that. And most will probably have to be smaller to allow room for the extra fuel needed to get the module into orbit. A load that large, even when partially empty, could add some 50,000 pounds (22,500 kg) to the shuttle's initial weight, requiring more energy to lift it off the ground.

Once in space the modules will all be attached to each other and also to a large metal framework that will resemble a giant Tinkertoy structure. The framework will be constructed in space from prefabricated pieces. This type of construction was proved possible by astronauts aboard the space shuttle in November 1985, and again by Russian cosmonauts in 1986. On the American mission, two spacewalking astronauts, while orbiting 230 miles (370 km) above the earth, easily snapped together 126 prefabricated pieces,

In this artist's rendering, two astronauts assemble sections of the space station out of the cargo bay of a space shuttle.

creating a 45-foot (14-m) tower. Of course, this was just a test of the final product, and therefore the astronauts didn't leave the structure out in space. They dismantled it and some other space-built structures that they had put together and brought all the pieces back down to earth, an exercise successfully completed.

ASSEMBLING A SPACE STATION

Each of the modules or sections that are brought up into orbit by the shuttle will at first be designed for a specific purpose. For example, a power system will supply the necessary electricity to run the station and the mechanism to keep the station in a desired orientation in space. The energy needed to run the station will come from solar panels and collectors that will capture the sun's radiation and convert it into the necessary electrical energy for the space station's operation.

Like the enormous wings of some prehistoric bird, these huge rectangular panels will extend way out beyond the main body of the station. From afar they will appear to dominate the entire structure, although they will consist of very thin sheets of a metallic substance called gallium arsenide. To test these winglike panels, the September 1984 shuttle *Discovery* deployed one while in orbit. The 102-foot (31-m) array of solar cells was unfolded from the cargo bay and, after proving that it could remain stable in space, was refolded and placed back into the bay.

One-third of the required electricity will come from this "photovoltaic" system. But making the solar panels larger so as to collect more radiation becomes difficult and so another system has been developed. Called the "solar dynamics generator system," it will use steam turbines. Solar heat will be collected and used to create the steam needed to turn the turbines. Together, both systems will produce 75 kilowatts of electricity. You could light one thousand 75-watt bulbs with that amount of energy.

Another section to be put into orbit by the space shuttle will store fuel, spare parts, and food and clothing for the astronauts. It will have to be positioned so as to be easily replaced every few months. Its entire replacement is felt to be more efficient than just resupplying it periodically with consumable items. A great deal of food and other supplies will be needed. During any ninety-day stay on the station, six people will require at least fifteen hundred meals—and that doesn't include snacks.

Water and oxygen, however, will be completely recycled in a "closed-loop life support system." Recycling is vital because of the large amount of water required. Every person must have six to eight pounds of drinkable water each day as well as thirty pounds of additional water for washing clothes and dishes and for personal hygiene. Full water recovery even from the humidity in the air is absolutely essential.

A third section necessary in the first stages of space station construction is the living quarters for the crew. The designs now being considered give each person his or her own personal room. These are not spacious places, however, and, in fact, will be more like large closets, a little bigger than a telephone booth. Within each cubicle will be a wall-mounted sleeping bag, an audiovisual entertainment center, places to store clothing and other personal items, and desk facilities. Remember that all will be weightless, and therefore the floor, walls, and ceiling can all be used interchangeably. That is why a sleeping bag can be located against a wall and why a chair is not necessary in front of the desk.

THE BIG TINKERTOY

An early space station construction base would probably have room for only six or eight astronauts, but it would soon be expanded as more modules and construction materials were brought to it by the shuttle. A combination workshop and laboratory similar in function to the Spacelab might be another such module. It would be hooked to the other sections in a predetermined location that allows the greatest efficiency. Future modules might be placed side by side;

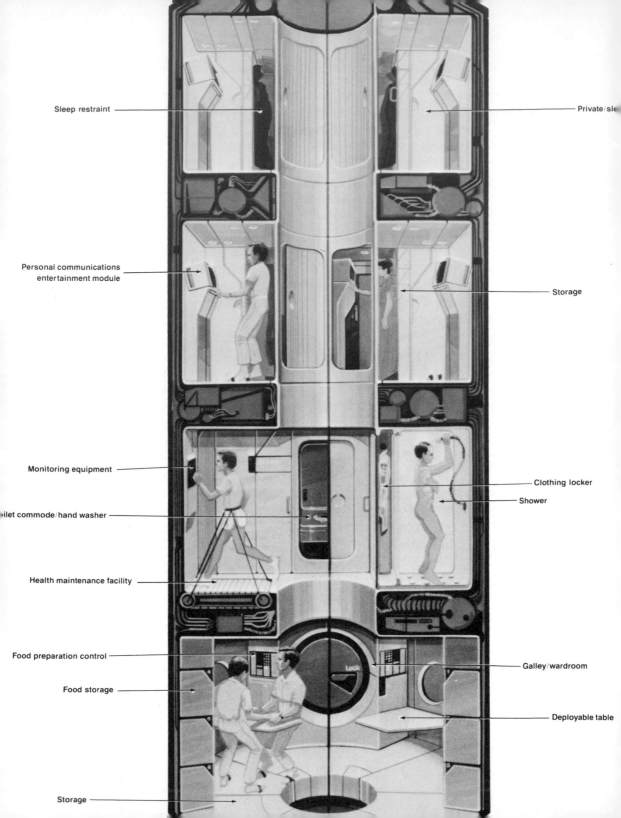

Sleep restraint

Private/sle

Personal communications entertainment module

Storage

Monitoring equipment

Clothing locker

Shower

ilet commode/hand washer

Health maintenance facility

Food preparation control

Galley/wardroom

Food storage

Deployable table

Storage

others may be attached at the end of modules already there. Once in space all the modules are weightless and therefore what might appear to be a crazy configuration here on earth may work extremely well in orbit.

Some of the later modules that are sent to the station will be special laboratories, each outfitted for the specific research projects it will house. One may be for biological space studies, another for manufacturing new alloys, a third for growing crystals in weightlessness. Some modules will be designed for storing materials. Close by will be workshops where these materials will be used to further the construction of the space station. Here the building of beams and girders, much like the tower made by the space shuttle astronauts earlier, will originate. The framework of girders will provide additional places for special modules and other instruments to be attached, as well as a place to install the solar panels and collectors that will convert solar energy into electricity.

Along with the basic space station, space platforms or "free flyers" will be constructed. They will not be attached to the station but instead will be capable of orbiting in predetermined paths. Although they will be unmanned, crew members clad in space suits will be able to visit these platforms to perform experimental and observational assignments. Some platforms will orbit near the station while others will be given polar orbits that will take them over the poles of the earth. They will be able to observe all parts of the earth from such a path and send back frequent reports on pollution or natural resources.

The space station will truly be an international venture even though NASA is responsible for its basic construction. The United States will contribute two modules, one for housing the crew and the other for hands-on laboratory work. The European Space Agen-

An early design concept for a 36-foot
(11-m) cylinder module for a space station.

cy, representing ten nations, will supply a life sciences laboratory, and Japan will construct a science lab devoted to advanced technology. The Canadians will make a remote manipulator arm similar to the one on the shuttle. It will be capable of moving about the station on a little trolley mounted on the girder superstructure. The United States and the European group will also each contribute two orbiting platforms, one that will stay with the station and the other that will be put into a polar orbit.

Once the solar panels are in place and the many different modules have been positioned together on the main structure, other secondary assemblies of antennas, industrial and scientific modules, and external servicing areas will be added. All of these will tend to give the station a false air of randomness. But before any additions are made to the basic structure, much planning will have taken place to insure efficiency and safety for all involved.

The addition of more living-quarter modules will provide room for a larger crew. How large the space station might ultimately grow depends upon its use. Some designers envision a station that would have several hundred people aboard at one time. They might not all be housed in the same complex of modules but would be able to work together if necessary. Eventually, there may be more than one space station in operation. At first, the inhabitants of these stations would be mainly astronauts, engineers, and scientists, but in time other occupations will be needed.

One early concept of a space station was aptly called a "power tower." One 400-foot-(122-m-)long girder was to have served as a large central spine. With that length, it would have been three and a half times as long as the space shuttle. Extending the entire length of the station, it would have had the interconnecting modules clustered at one end. Another girder at right angles to the central spine would then have held the solar panels. In other places along the main girder would be experimental stations that could be operated by remote control or by astronauts in space suits. Satellite launchings and repair services also could have been managed from such locations.

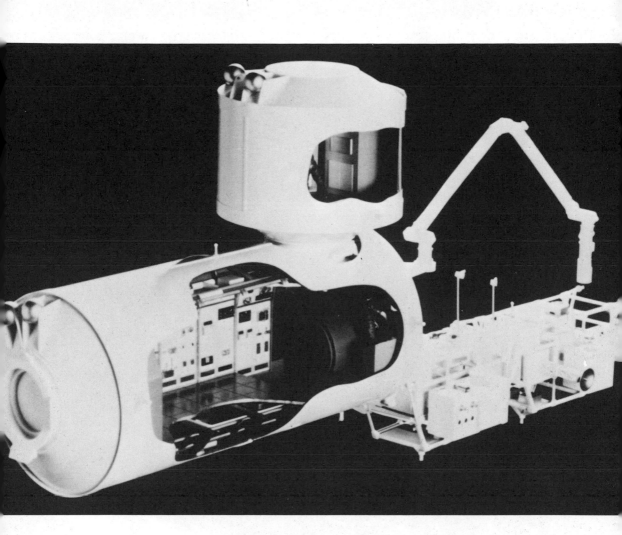

A detailed model showing the interior of a Japanese experiment module being studied by Japan as a potential contribution to NASA's permanently manned space station. The model features a large, pressurized laboratory module, an exposed work deck, and a local remote manipulator arm. The smaller module on the top is an experiment logistics module.

The power tower design was replaced in 1986 by what is considered a more rigid structure. It also will provide more room on which to install instruments and other experimental stations. Instead of having one long spine, the new model will have two such structures, each 297 feet (91 m) long. They will be connected by other girders to form a rectangular configuration. This is called the "dual keel" model. Like a boat with two keels, it will move in what President Kennedy once called the new ocean of space.

In choosing the dual keel model over the power tower, a little stability has been lost, and the new model will require more energy to control its orientation in space. In the power tower model the difference in the amount of gravitational pull from one end of the tower to the other would have caused it to align itself naturally with one end pointing toward earth. Not only would this have helped to eliminate any wobbling of the space station, but it would have also allowed earth-observing instruments to be placed at the end facing earth and astronomical instruments at the other end. Fatter configurations, such as the dual keel, don't have such stability, but the other advantages of the dual keel model are felt to outweigh the slight loss of stability.

PERMANENT STATION, TEMPORARY CREW

Although the space station itself is designed to be a permanent structure, the personnel aboard it will only stay in space for a limited amount of time, probably for no more than three to six months. Unlike Skylab, however, the station will never be completely deserted. Instead, a rotating crew of highly trained astronauts will maintain it, while scientists, engineers, and others who have special

One early concept of a space station was called a "power tower." It featured a 400-foot- (122-m-) long girder which was to serve as a large central spine.

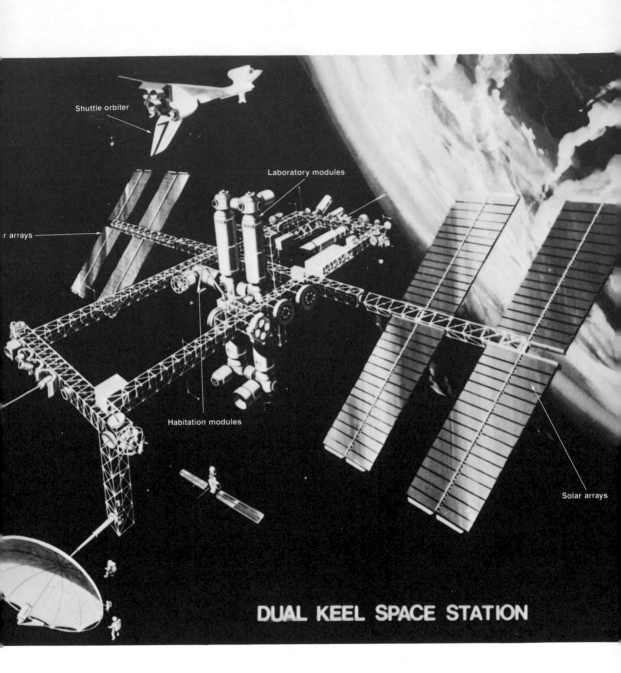

Shuttle orbiter

Laboratory modules

r arrays

Habitation modules

Solar arrays

DUAL KEEL SPACE STATION

projects will spend whatever time is necessary aboard the station.

This constantly changing crew is necessary, first, because of the effects of weightlessness on the human body over a period of time. As has been shown by the crews of Skylab and Salyut, little harm is done in short periods of weightlessness, but there are measurable changes. None of these effects have been found to be permanent, even after two Russian cosmonauts stayed in space for 237 days. But we don't know what would happen to a person after more than a year in weightlessness. Before anyone is allowed to remain in space that long, test animals will be kept in the space station and watched carefully.

Another reason for rotating the space-station crews is that not many people are ready to leave their homes and families for such long periods of time. Eventually, when the space station has evolved into a much more sophisticated structure, many envision the possibility of entire families living and working for many years out in space. But that development is a long time in the future and would not be considered simply a "station." In fact, the space stations that are now being planned may very well be the stopping-off places on the way to such a future dream.

The "dual keel" concept of a space station features two approximately 297-foot- (91-m-) long vertical keels which provide added space where payloads can be attached, and place the pressurized modules near the station's center of gravity to provide more volume for experiments designed to take advantage of the near-zero gravity of space. Nearby is a co-orbiting platform. At the top of the picture, a space shuttle orbiter prepares to dock at the space station.

CHAPTER SIX

ABOARD THE SPACE STATION

A sunrise seen from an orbiting spacecraft like the shuttle is very different from any you have ever seen from earth. First, because the shuttle moves so much faster than the rotation of the earth, the sun appears to rise much faster than what we are accustomed to. There isn't the prolonged and gradual predawn light that grows brighter and brighter until the sun finally appears above the horizon. Instead you first see varying bands of different colors before the sun looms suddenly into view. These bands range from brilliant reds to the brightest, deepest blues and never seem to be the same from one sunrise to the next. The astronauts aboard the Shuttle can watch sixteen sunrises and sunsets every twenty-four-hour period, but even with that many of them to observe every day, those who have seen them never seem to get tired of watching for the next one.

But there is usually little time to spend watching sunrises. Idleness is rare in space. Many astronauts, in fact, complain that their work load is too heavy and they don't have enough time to finish all their assignments. Some of the procedures in conducting an experiment, for example, may require following very precise instructions and may take much longer than the time allotted for them by Mission Control down on earth. And simple tasks that we take for granted at home, such as preparing food, taking a shower, or cleaning up, are not routine jobs in space. They require great care if they are to be done successfully. Crumbs of food or droplets of water floating around the cabin can get lodged in unwanted places and possibly cause malfunction of some important instrument or mechanism.

CREATING A
LIVABLE ENVIRONMENT

The experiences and suggestions of the many astronauts and cosmonauts who have spent various amounts of time in space have been taken into consideration by the designers of the space station. Although it is not a reality yet, enough plans and proposals have been presented to enable us to get some idea of what it will be like living in such a craft. Because missions of three to six months are being planned, even the early facilities will have to be able to support people reasonably comfortably for such time periods.

As was mentioned earlier, the different sections will be brought up into space by the shuttle. Current plans call for at least nine shuttle flights before the space station will be permanently manned and in operation. The module that will house the crew is scheduled to go up last, so that when the first crew arrives to set up housekeeping, they will be able to commence doing useful work immediately. However, it will take another nine or ten shuttle flights to complete the station, including constructing the platforms and the modules being contributed by other countries.

Once the space station is big enough to house more than just the initial six or eight astronauts, many different activities will be going on at all times. And not everyone will return to earth at the same time. Some may stay for two three-month periods, so that although the crew will be rotated, the station will never be left completely empty.

Also, in order to keep the facility operating continuously, crew members will be divided into alternate shifts during any three-month period. Many factories on earth do this, but here it means that some people have to work at night and sleep during the daytime. In a space station, on the other hand, with sixteen sunrises and sunsets every twenty-four hours, time cannot be determined by whether it is daylight or night outside. Everyone aboard will experience frequent day and night periods during any time they are awake and active. Therefore, as with earlier earth-orbiting manned spacecraft, the space station clocks will be set according to some time zone on earth.

What would the daily routine of a crew member be like? It is estimated that everyone will put in six hours per day on his or her payload work and spend about two hours per day on housekeeping. They will have to devote at least two hours per day to exercise in order to keep their muscle tone and avoid calcium loss in their bones. They will also be given one or two days off every week.

On earlier space missions, both the Americans and Russians found that any heavier work schedule produced negative results. Longer work days tend to produce more mistakes and less efficiency. People get tired and bored more quickly and "burn out" if pushed beyond certain limits. On the other hand, by changing people's schedules, providing more variety in the work itself, and allowing more frequent contact with families and friends on earth, crew members actually tend to improve and become more expert in their assigned tasks. Other factors that affect work production and must therefore be considered in space station design and functioning are the need for some privacy, comfortable places to relax when off duty, and good, nourishing food.

REST AND RECREATION

Each person aboard the space station will have his or her own personal cubicle. As we saw in the last chapter, these are not spacious but they do allow crew members some privacy, which was lacking on previous craft. Also, everyone will be able to adjust the temperature, lighting, and ventilation within his or her own living quarters. They may even be able to rearrange the walls and change the shape of the rooms to suit their own comfort. And it is unlikely that many will spend very much time within these rooms because of the busy work schedule planned for them. The cubicles are being designed primarily for resting and sleeping.

During their time off, when they aren't sleeping or otherwise occupied within their own rooms, crew members will have various activities available to them in a multipurpose wardroom. There they will be able to play games with other off-duty personnel, possibly have a snack to eat, or just relax after a hard day's work. The exercise machines will also be placed in the multipurpose room, and it will be mandatory that everyone use them at least two hours per day to counteract the effects of the weightless environment. To relieve the boredom of this activity, television and radio will also be part of this recreation room.

Nearby, within the same module as the wardroom, will be the galley in which meals will be available. All of the latest conveniences will be there, such as a refrigerator, dishwasher, trash compactor, and possibly even a microwave oven. Preparation and cleanup will be made as simply and quickly as possible. The food itself will be similar to that offered to space-shuttle crews, with perhaps a greater selection from which to choose. One of the hardest problems spacecraft designers must grapple with is the tedium of long periods of space travel. The ability to choose from a large variety of foods is one way to avoid this. The Skylab astronauts, for example, suggested that more steaks and ice cream be provided because these foods were the most popular and were depleted first. The Russian cosmonauts were found to eat a lot of sweets at first, such as cookies and candies, but as their tastes changed, they later preferred salty and spicy things.

Within the galley area there will also be the ship's laundry. Without such a facility, clean clothing and linen would have to be periodically brought up to the station by the space shuttle and stored until ready for use. Large storage areas would be needed for this purpose, and because space is at a great premium in any spacecraft, washing and drying machines are much more practical. At the beginning, everyone will probably be responsible for his or her own laundry needs, but once the space station grows larger, specific people may be in charge of the laundry facility.

WORKING IN SPACE

The major occupation of the crews of the space station will be scientific and engineering projects. These will occupy most of their time and probably dominate most of their conversations and thoughts. Some of the engineers will be busy building additions to the existing structure so that more people and experimental places can be accommodated. They may also be building new structures that will eventually become the framework for a second or third space station. The scientists will be conducting experiments to determine, for example, how fluids, metal, and gases, as well as humans and other animals, function in zero gravity. They will also be designing new space products.

The space-station engineers will be the ones to maintain and repair the unmanned space platforms or "free flyers" which will orbit the earth separate from the station. Then, once these platforms are released into their own orbits, other personnel from the station will be able to use them to conduct specific experiments, observations, or other activities. These platforms may house astronomical instruments too sensitive to tolerate the slight motions of the space station. Or they may hold dangerous volatile materials.

The frequent arrivals and departures of personnel to and from the platforms, as well as the quarterly visit of the space shuttle, will require crew members skilled in handling rendezvous and docking procedures. It would not take too great a push by an incoming craft to destabilize the entire station, causing it to develop a wobble and possibly ruin or damage some of the delicate instrumentation that will be aboard it.

Other people aboard the space station will be working in the laboratories, doing experiments, making observations, creating new synthetic materials, or manufacturing special products. They will all be taking advantage of the weightless environment and,

This artist's concept of a space station focuses on the pressurized modules where the crew will work and live. Two members of the crew inside the cupola atop the right-hand resource node control the remote manipulator arm. In the background, a co-orbiting platform flies in tandem with the manned base. An orbital maneuvering vehicle is shown flying out toward the platform.

especially for the astronomers, the pollution-free, airless vistas of deep space. Geologists will make observations of the earth, and ecologists will determine where mining or agricultural activities would be most profitable. A broad, space-based communications program with maintenance and upgrading of communication satellites will be feasible.

Besides the studies, experiments, and observations that deal with terrestrial activities, some personnel will be involved with the future planning and construction of possible bases of operation on the moon or on one of the planets. The space station will be the base for expeditions to the rest of the solar system. We might be able to set up mining operations on some of the asteroids or find other useful raw materials on the moons of the other planets. With a space station base, getting to these distant places will be made much easier. The power needed to launch a spacecraft from a base that is already in orbit around the earth is a great deal less than the power needed to lift it off the earth.

Two other future concepts are also being considered, both of which will use the space station as a launching platform. A space colony has long held a fascination for many people, and organizations are even evolving plans for such a structure, although it will be many years before the building of one can even be considered. And, whereas a space colony will be put into orbit around the earth, some dreamers envision a spaceship that will travel out to the distant stars, possibly not returning to the earth for centuries. Like the establishment of a base on a solar-system body other than the earth, each of these will also require a great deal of research, planning, and careful construction. The money, time, and work required will be enormous but ultimately worthwhile for all humankind.

CHAPTER SEVEN
SPACE COLONIES

A space colony, until very recently, was science fiction material. Few scientists or engineers took the idea seriously. But today many foresee the time when such a colony will become a reality. The seriousness with which this concept has been considered was seen in the many conferences and seminars held in the mid-1970s addressing this issue. These meetings brought together physicists, astronauts, spaceflight technologists, and other interested parties to study the feasibility of building and inhabiting a space colony or perhaps even many colonies.

Although many ideas and plans will change, just as they did for earlier stages of space exploration, it is fairly certain that eventually human beings will be able to live permanently in specially constructed space colonies. Unlike the space station in which a person can stay for only a few months, the space colonies will house entire families. They will be small communities with their own governments, service facilities, recreational activities, schools, and stores. Thousands of people will live in each colony. Their reasons for leaving the colony will be similar to those given by earthlings today—vacations, business, exploration of new regions, etc.

The space shuttle, as we have seen, will be the vehicle to bring the parts of the space station into orbit around the earth. The shuttle, however, was only designed for earth orbit and can't go beyond that region. Other vehicles will have to be built to go from the station to the moon, the planets, or a space colony. Once the space station is completed and operational, it will truly become a station, a stopping-off place on the way to more distant bases.

MINING THE MOON

One of these bases will be on the moon. The United States's Apollo missions to the moon revealed that the moon's crust is rich in aluminum and titanium ore. It also contains silicon and oxygen. All of these raw materials can and will be used in the construction of a space colony. Because the moon has only one-sixth the gravity of the earth, the lifting of these materials from the surface of the moon will require much less energy than from the earth.

Once the rich soil is extracted from the lunar surface, it will be put into specially designed buckets and catapulted by a "mass driver" across hundreds of thousands of miles of space to an orbiting space station. The station will have a "catcher" that retrieves the space-flung buckets, empties them, and sends them back to the moon for refilling. The space station will also house facilities to begin processing the lunar ore, although much of that work will be done at the colony's construction site. Special space tugboats will deliver the usable materials from the space station to the colony.

This two-stage delivery system will continue even after the colony is established. By that time there will be a great deal of traffic in and around the colony, and the stream of buckets carrying ore might dangerously interfere with the spacecraft coming and going. Also, the space station that receives the buckets can be located closer to the moon, allowing a smaller catcher or a less accurate mass-driver to be used.

The lunar mining base may eventually grow into a permanent community itself, but it will differ from the one that develops on the space colony or from any terrestrial community. These differences will arise not only from each community's location but also from the type of people who will settle them, the resources available to them, and their leadership. All of these factors helped shape the various characteristics of the early European settlements in America, the new world of the sixteenth and seventeenth centuries. And colonies in space or on the moon will be the twenty-first century's new world.

SELECTING A SITE
FOR A COLONY

Where will the first space colony be located? It must be in a place that can be reached within a reasonable amount of time by those who wish to go there. And, because so much of its raw material will come from the moon, it would be advantageous to locate it near the moon as well. It is also important that the colony's location be a stable one so that a minimum amount of energy would be needed to keep it in place. An orbiting object too close to either the moon or the earth would eventually be pulled out of its orbit and, if its path was not quickly corrected, it would crash into whichever body was tugging on it the hardest. An orbit is needed in which the gravitational pull of both the earth and moon balance each other.

In the eighteenth century, the French mathematician Joseph Louis Lagrange calculated five moving points in space where this gravitational balance between two bodies can be achieved. Any object at one of these points would tend to stay there rather than fall toward either body. These five positions are known as "Lagrangian libration points." Because the two bodies, in this case the earth and moon, are in constant motion, the Lagrangian points are also constantly changing their locations in space. But they always keep their same relationship with the positions of the two bodies.

Although there are five Lagrangian points, two of them are the most stable ones. These two are known as L4 and L5. For the earth-moon system, both are located in the moon's orbit, but L4 lies ahead of the moon and L5 behind the moon. Both are 240,000 miles (386,000 km) from the moon, which is also the distance of the earth from the moon. As you can see in the diagram on page 62, the Lagrangian points L4 and L5 each make an equilateral triangle with the earth and the moon.

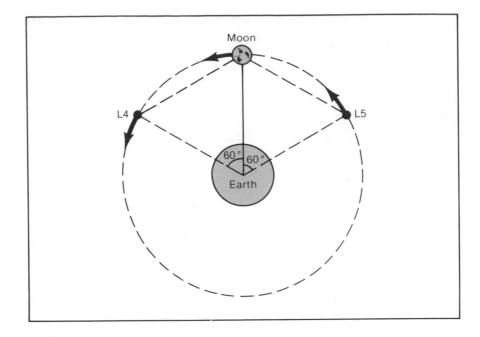

The two stable Lagrangian points, L4 and L5.
Note that both form an equilateral
triangle with the earth and moon.

The first space colony probably will be located at L5 and will orbit the earth by following the moon in its orbit. The choice of L5 rather than L4 will be an arbitrary one, however, and perhaps a second colony will be built at a later date at L4. Both colonies will maintain their moving positions in space in the same way that two groups of asteroids lead and follow the planet Jupiter in its orbit. These are called the Trojan asteroids. They are located in the Lagrangian points L4 and L5 in Jupiter's orbit. The asteroids remain equidistant from Jupiter and the sun at all times and are therefore able to maintain their positions without being pulled closer to either Jupiter or the sun.

SIMULATING GRAVITY

The usual picture of a space colony is that of an enormous wheel with a central hub. When the colony is actually being built, however, other shapes may prove to be better. Regardless of what form the colony takes, the section in which people will live must be made to rotate. This will create an artificial gravity and thus avoid the problems created by long periods of weightlessness.

Perhaps you have gone on an amusement park ride which rotates very fast, much faster than the usual merry-go-round. In this ride everyone stands on the inside of a high circular wall. The entire unit, including the wall, rotates, and as it builds up speed you feel yourself being pushed with greater and greater force against the wall. At full speed your feet may no longer be touching the floor, but instead you are securely pinned against the wall by the force of the whirling unit. If you close your eyes to shut out the whirling scenery, you will feel as though you were lying on your back rather than standing erect. The circular motion of the ride has created a feeling of gravity that is greater than the real pull of the earth's gravity. Of course, it isn't real gravity that you are feeling during the ride, but it has the same effect nevertheless.

The rotation of the space colony will have a similar effect upon its inhabitants. Its rotational speed does not have to be as great as the amusement-park ride since it only has to simulate earth gravity, not overcome it. It has been calculated that a colony with a diameter of about a mile would require one rotation per minute to create an artificial gravity similar to the real gravity on earth. However, an earthlike gravity will only be felt in the outermost regions of the rotating colony. As you approach the center of the structure, less and less gravity will be felt. And because of this, you will think of the direction "down" as being toward the outside of the colony where the gravity is the greatest. "Up" will be toward the center.

In fact, if you live on the upper floor of an apartment building in the space colony, the gravitational effect on you will not be as great as it is on the ground-floor inhabitants. This is also true on earth, but the difference in gravity between two floors in a building here is so small as to be unnoticeable. The top and bottom floors of even the tallest buildings on earth are separated by a very small amount compared to the size of the earth. That ratio will be much greater in a space colony, and the effect may be quite obvious. Whether this will present problems for the future inhabitants of the colony remains to be seen.

DESIGNING THE STRUCTURE

What will the colony look like? Let me describe one possible model. Imagine a 400-foot (122-m) wide hollow metal doughnut with a circumference of about 4 miles (6 km). At the center of the doughnut is a central hub about 400 feet in diameter. It is connected to the doughnut ring by a series of large spokes. Because "down" is toward the outer rim of the doughnut and "up" is toward the hub, elevators located in the spokes will take people back and forth ("up" and "down") to different parts of the colony. The spokes are 50 feet (15 m) wide and therefore also have room for cables and pipes as well as offices, stores, and scientific laboratories. They will resemble high rises here on earth, although it will take some adjustment to ride up in an elevator, float through an almost weightless area, and then take another elevator down to your destination on the other side of the colony.

The rim of the space colony is where everyone will live. There you will find homes, parks, playgrounds, ballfields, and gardens, as well as terraced plots of ground for growing food and raising cattle. Sunlight will come in through windows located on the inner one-third (the "top") of the doughnut. A huge mirror suspended above the entire structure will capture the solar radiation and reflect it onto the colony. Because the sun's light will be regulated, as will any water sprinkling on the crops, there won't be any major changes in weather conditions. Snow, torrential rains, hurricanes, and tornadoes will be nonexistent on the space colony.

A space colony might look like a giant doughnut. The central hub is connected to the outer ring by a series of large spokes (top). The rim of the space colony is where everyone will live. A giant mirror suspended above the structure will capture the solar radiation and reflect it onto the colony (bottom).

The outer rim of the doughnut will be made of thick aluminum wrapped in cable for shielding against radiation. The earth's atmosphere may cause weather problems, but it also protects us from powerful and deadly rays from the sun and from outer space. Excess radiation may prove to be one of the biggest problems to be dealt with in designing a livable space colony. It may be possible to surround the ring with a thick layer of moon rocks and soil. Or, as some planners have suggested, a subway system built below the housing and agricultural areas would help insulate the living areas from radiation. That would also provide another way of traveling around the colony. You would not always have to go through the changing gravities to reach another place.

WHAT ARE THEY DOING UP THERE?

Ten thousand people will be able to live in this proposed space colony. They will be permanent inhabitants although they will still be in contact with the earth and with any communities on the moon or elsewhere in the solar system. They will see the same movies and read the same books that are available to earthbound people. They will be able to telephone their families and friends back on earth, and mail service will provide further contact between these distant places. The children in the colony will attend school and eventually may even be able to go to a space college.

One of the colony's principal businesses will be the construction of large satellite power stations. These will collect sunlight, convert it into electricity, and send it in microwave form to stations on earth. There the microwave energy will be changed back into usable electricity. The power stations will become a vital source of energy for the earth. The colony will also be able to construct additional colonies, some of which may be much bigger than the initial one. One study designed a spherical colony with 250 square miles (648 sq km) of living area. That's almost as big as the city of Chicago.

Other economic activities would include general manufacturing. Some processes can be done more efficiently and less expensively in space. And many products, if produced under weightless conditions, are far superior to any made on earth. For example, as already proven aboard manned spacecraft, it is possible to produce the purest pharmaceuticals, the most sensitive electronic parts, and the strongest metal alloys ever made when zero gravity and a high degree of vacuum is available. Although humans will not be able to leave the colony without spacesuits or pressurized space vehicles, manufacturing in the vacuum of space can be accomplished with computers and remote control devices as well as brief check-up visits by scientists and engineers.

Many problems still need to be solved before a space colony can become a reality. But it will not lack for settlers. Already there is an organization called the "L5 Society" whose members want to be the first settlers in space. Whether these people will be among the first to be chosen as citizens of space will depend upon the criteria used for selection. In the beginning, at least, those with high intelligence, the necessary training, and good physical and mental health will probably be given priority. In later stages, once the colony is established, it may be open to just about everyone. Would you like to live there?

CHAPTER EIGHT
JOURNEY TO THE STARS

The lure of space will call some people to explore farther than the boundaries of the solar system. Even before the space colony is fully operational, some will be ready and eager to head for the stars. Just as the planners of the space colony have already started to draw designs, hold conferences, and test some of the proposed ideas, so deep-space enthusiasts will start planning long before such an expedition is feasible. And of course science-fiction writers have been describing interstellar travel in great detail for many years.

The space colony will probably be the ideal place to build a spaceship. First of all, the engineers and scientists there can draw upon their experiences in building not only the colony itself but also many smaller craft that will already be traveling within the solar system, visiting outposts and colonies on the planets, moons, or asteroids. Also, the spaceship builders can take advantage of weightless conditions when they transport raw and processed materials to the building site. By then perhaps much stronger materials will have been developed that can make the ship more durable and safer.

HOW FAR IS FAR?

A spaceship headed beyond the solar system will probably have to be bigger than any existing craft because its journey will last at least several years. And even in that time, no craft can hope to reach even the next nearest star. To understand why this is so, it would be helpful to scale down space distances to units that we are more familiar with. The distance between the earth and sun is 93 million miles (150 million km). If we use a scale of 1 inch equals 93 million miles (2.54 cm = 150 million km), the earth will be located 1 inch (2.54 cm) from the sun. The giant planet Jupiter will be 5 inches (12.5 cm) from the sun, and the most distant planet, Pluto,* will be 39 inches (99 cm) away. On that same scale, however, the star nearest to our sun will be over 4 miles (6 km) away! That star is called Alpha Centauri.**

Because of the enormous distances between the stars, astronomers don't use miles or kilometers in their measurements. They use "light years." A light year is the distance that light travels in one year. Since a ray of light travels 186,000 miles (299,460 km) per second, by computing the number of seconds in a year and multiplying that figure by 186,000 we find that a light year is approximately 6 trillion miles (9.7 trillion km) long. Alpha Centauri is in reality about 25 trillion miles (40 trillion km) away. Its light takes four and one-third years to reach us.

MORE THAN A LIFETIME

Since no spaceship can travel even close to the speed of light, to reach the stars will require many more years than the lifetime of any human being. Anyone embarking on such a journey must be willing to sever all contact with the earth forever. The children or grandchildren of such pioneers may one day return to earth, but only if the spacecraft is self-sufficient enough to support future generations.

---

*Although Pluto is currently closer to the sun than Neptune, its average distance from the sun is much farther away than Neptune's average distance. Pluto is therefore considered the farthest known planet in the solar system.

**Alpha Centauri is actually a member of a small star group that includes the star called Proxima Centauri. Proxima Centauri is the closest star to us at the present time, but astronomers usually refer collectively to this star system as Alpha Centauri.

Obviously, any spaceship sent to the stars must be far more independent than the space colony. Once out of radio contact with the earth, the space travelers are on their own.

As we have seen, to escape the earth's gravity an object must be propelled upward at the speed of 7 miles (11 km) per second. Assuming that a spacecraft could be sent out of the solar system at that velocity, it would still take it well over one hundred thousand years to reach the vicinity of Alpha Centauri. Even if the craft could travel at ten percent the speed of light or 18,600 miles (30,000 km) per second, a velocity that no artificial object has reached or is likely to in the near future, Alpha Centauri would still be at least one hundred years away. And of course another one hundred years would be required to return to earth.

Assuming the safe return of the spacecraft after those two hundred years or so, what would the crew find here on earth? Although they would not be the original people who left the earth, they would have learned about it from their parents. They would, however, encounter a very different world than the one that their parents or grandparents left. Just think of how much life has changed in the last fifty or one hundred years. Although no one can predict how fast or slow future changes will occur, we know that the world will be quite different in the year 2040 than it is now. Can you imagine what it might be like in the year 2240?

To many space enthusiasts, returning to the solar system is not necessary. These people envision building space colonies that are capable of traveling out into space, never to return. The craft would be similar to the colonies described in the last chapter but instead of remaining in orbit close to the earth, it would leave the solar system. Many designs already exist for propulsion systems that would enable these "colony ships" to travel to the stars, and for some space enthusiasts it really wouldn't matter if it took five generations or more to find another region to colonize.

Once a star with orbiting planets is reached, the colony ship, like the ships that explored new lands centuries ago, will "weigh anchor." The explorers will need energy, which they will obtain from the star, and raw materials, which they will mine from the planets and moons. They will build new space colonies and eventually colonize the inhabitable planets as well. Once these settlements are well established, they in turn will send out their own starships to more distant stars. In this way, it is believed, the human race will some day be spread out over a large region of space. Perhaps the entire galaxy, instead of just one small planet, will be our home.

FOR FURTHER READING

Bendick, Jeanne. *Space Travel*. New York: Franklin Watts, 1982.

Hawkes, Nigel. *Space Shuttle*. New York: Franklin Watts, 1983.

Kerrod, Robin. *See Inside a Space Station*. New York: Franklin Watts, 1979.

Ride, Sally, with Susan Okie. *To Space and Back*. New York: Lothrop, Lee, 1986.

INDEX

910.0919 Apfel, Necia H.
APF
        Space station

$10.40   2688

© THE BAKER & TAYLOR CO.